Talkmore Ziteya

The Economics of goat milk production

Talkmore Ziteya

The Economics of goat milk production

Socio-economic factors influencing goat milk production in the smallholder areas of Zimbabwe

LAP LAMBERT Academic Publishing

Impressum / Imprint

Bibliografische Information der Deutschen Nationalbibliothek: Die Deutsche Nationalbibliothek verzeichnet diese Publikation in der Deutschen Nationalbibliografie; detaillierte bibliografische Daten sind im Internet über http://dnb.d-nb.de abrufbar.
Alle in diesem Buch genannten Marken und Produktnamen unterliegen warenzeichen-, marken- oder patentrechtlichem Schutz bzw. sind Warenzeichen oder eingetragene Warenzeichen der jeweiligen Inhaber. Die Wiedergabe von Marken, Produktnamen, Gebrauchsnamen, Handelsnamen, Warenbezeichnungen u.s.w. in diesem Werk berechtigt auch ohne besondere Kennzeichnung nicht zu der Annahme, dass solche Namen im Sinne der Warenzeichen- und Markenschutzgesetzgebung als frei zu betrachten wären und daher von jedermann benutzt werden dürften.

Bibliographic information published by the Deutsche Nationalbibliothek: The Deutsche Nationalbibliothek lists this publication in the Deutsche Nationalbibliografie; detailed bibliographic data are available in the Internet at http://dnb.d-nb.de.
Any brand names and product names mentioned in this book are subject to trademark, brand or patent protection and are trademarks or registered trademarks of their respective holders. The use of brand names, product names, common names, trade names, product descriptions etc. even without a particular marking in this work is in no way to be construed to mean that such names may be regarded as unrestricted in respect of trademark and brand protection legislation and could thus be used by anyone.

Coverbild / Cover image: www.ingimage.com

Verlag / Publisher:
LAP LAMBERT Academic Publishing
ist ein Imprint der / is a trademark of
OmniScriptum GmbH & Co. KG
Heinrich-Böcking-Str. 6-8, 66121 Saarbrücken, Deutschland / Germany
Email: info@lap-publishing.com

Herstellung: siehe letzte Seite /
Printed at: see last page
ISBN: 978-3-659-79805-4

Zugl. / Approved by: Harare, University of Zimbabwe, Diss., 2013

DEDICATIONS

This research project is dedicated to my special mother and mentor, Mrs. Barbra Ziteya Mangwende and the entire Mangwende family for their sincere love and commitment to my studies. It was a sacrifice on their part. Great deeds need great preparations and the reward of a thing well done is to have done it. I am grateful to everyone who did something in order for me to achieve my academic dreams. Just like a field cannot bloom if there is only sunshine and no rain, I would not have done it if you guys were not supportive. May the good Lord richly bless you in abundance.

ACKNOWLEDGEMENTS

Above all, I would like to thank the almighty God for without his grace this research would not have taken place. My sincere gratitude goes to the University of Zimbabwe for giving me the opportunity to pursue my first degree. Moreover I appreciate the contribution made by the entire staff of the Department of Agricultural Economics in my academic studies. I would also want to show my appreciation to Mr. Chiwashira. Special thanks go to my supervisor, Mr. T. Chamboko for his great contribution in my research project. His valuable guidance, support and dedication have made it possible for me to accomplish the project. I am also indebted to Department of Agricultural Extension Services in Plumtree especially Mr. Njabulo Hanyani, Mr. Bekhumuzi Mpofu and Mr. Given Masendeke for their support during data collection.

TABLE OF CONTENTS

LIST OF TABLES   **PAGE**

LIST OF FIGURES   **PAGE**

ACRONYMS

FAO	Food and Agriculture Organization
HPIK	Heifer Project International-Kenya
ICRISAT	International Crops Research Institute for the
	Semi-Arid Tropics
AREX	Agricultural and Extension Services
AGRITEX	Agricultural Technical Extension Services
ORAP	Organization of Rural Association for Progress
NVS	Netherlands Development Organization
SPSS	Statistical Package for Social Scientists

CHAPTER 1: INTRODUCTION

1.1 Background

The estimated world population of goats is over 924 million (FAOSTAT 2011). Countries in Europe and areas around the Mediterranean region have the most developed goat industry and dairy goat focused research. The Mediterranean area is the main goat milk and goat cheese producer (18%) outside of India (22%), which has the greatest goat milk volume of all countries, but keeps mainly only dual purpose goats (meat and milk) (Dubeuf et al, 2004).

In developing countries for instance in Zimbabwe, goats are sometimes viewed as the poor peoples' animal and are primarily owned by smallholder farmers. Zimbabwe's small stock population comprises of 3.3 million goats and 390000 sheep. Approximately 97% of the goats are owned by smallholder farmers (van Rooyen et al, 2007). The majority of goat breeds in Zimbabwe are indigenous. The goats are either of the smaller type (East African goat) found in the eastern and central areas or the larger type (Matebele goat) of southern and western Zimbabwe, although there have been efforts to introduce improved breed such as the Angora goats. .

Goats contribute significantly to the livelihoods of smallholder farmers in Zimbabwe. Smallholder farmers rear goats for meat, milk, skin and manure production. They also rely on goats for income generation. Zimbabwean goats are also important for food security, and they act as a form of insurance during periods of crop failure. Goats are also slaughtered during religious and customary rites as well as festive occasions. Blood and bone meal from goats also have a commercial value and can be sold to generate income for the smallholder farmers.

Goat products seldom enter the formal marketing system and so the goat's contribution to the rural and national economy tends to be grossly underestimated (Peacock, 1996). Goats are usually kept by poorer people, often tended by woman, who seldom have a voice in national discussions. Due to these reasons, goats and people who keep them are accorded a low status and given a low priority in national development.

The most publicity received by goats has been about their undeserved reputation as environment degraders. The support to livestock itself was biased towards commercial cattle production. Goat production was not encouraged and received little research and development (Ndlovu 1994).

This policy was, however, reversed after independence in 1980 (Sibanda and Khombe 2006). Despite this change in policy, like many developing countries, the negative image of goat dairy products and a lack of defined markets for goat milk have long been recognized as a handicap for organizing and appreciating an effective dairy goat sector (Debeuf et al 2004).This is also despite the fact that the composition of goats' milk, for example, has been found to differ from cow or human milk in having better digestibility, alkalinity, buffering capacity and certain therapeutic values in medicine and human nutrition (Yangilar 2013).

Largely as a result of prejudice and ignorance of the importance of goats to farmers in rural areas there had been little research on goats in developing countries until about 20 years ago (Peacock, 1996). Scientists in a few developed countries carried out research to support the intensive dairy goat industries of Europe and North America, but little research was done on goats in developing countries. Over the last 20 years a considerable body of knowledge has accumulated on this previously neglected species.

 In Zimbabwe some initial research work was carried out at Matopos Research Station as early as 1975 but comprehensive studies started after independence (Lindela et al, 1991). In 2011, Christian Aid carried out a ZimPro/ LDS farming project in Manama communal area in Gwanda. The goat project had an aim of improving incomes and livestock distribution. The project had an aim of helping 380 farmers but due to low livestock prices in Gwanda the project was able to buy more animals for 530 families. Each household received three goats or sheep and three chickens or guinea fowls. They also received training in livestock production, control of diseases, housing, and good feeding practices. The veterinary department vaccinated all the animals. Livestock committees were set up in each of the five areas where animals were handed out so as improve the sustainability of the project. The livestock committees were responsible for monitoring the welfare of the animals and negotiating for good livestock prices. A scheme was set up so that when these animals have offspring, they were passed on to other families in need thereby multiplying the benefits.

Feeding goats well is of fundamental importance for the success of the whole goat enterprise. Good nutrition is a prerequisite for good health, good reproduction, high milk yield, fast growth rates and a successful goat system (Peacock, 1996). Studies on goat nutrition have included diet selection on range, evaluation of range and browse species, forage legume supplements and the use of high energy fattening diets for growing goats. These studies have contributed to the growing body of knowledge on goat production.

Goat's milk is used for drinking; making cheese and yoghurt. The milk is very important in the smallholder areas because it improves the diet of rural people. It is traditionally valued for the elderly, the sick, babies, children who are allergic to cow's milk, and patients with ulcers (Harlan et al, 2003). Goat's milk is richer than cow's milk in nutrients such as Vitamin A, niacin, choline and Inosil (Kosgey et al, 2006). There is an increasing interest in goat milk production, and there are now goat milk development programs in many countries (Morand-Fehr et al, 1991).

In the tropics goats are rarely kept just for milk but goat's milk and milk products are of great importance in subsistence agriculture (Steele, 1996). The milk is recommended to HIV patients because its high proteins molecules are better absorbed than other proteins therefore they are known to strengthen antibodies. For mothers who opt not to breastfeed the milk is an ample substitute. Goat's milk is highly nutritious and has a similar nutritional profile to human milk, containing 4.5% fat, 4.0% lactose and 3-4 % protein depending on the goat's nutrition, breed and stage of lactation (Peacock, 1996). Onim (1992) found out that the contribution of goat milk to the improvement of protein poor diets of small-scale farmers is very important especially for children.

Pure exotic or crossbred dairy goats and associated technologies are preferred as a fast means of improving animal production of smallholder farmers and quickly enhance their economic status and diet quality (Kosgey et al, 2006). Since goats are more resistant to draughts they make use of a wider variety of plants and their higher reproduction rate allows populations to recover quickly. As browsers they can also use different vegetation compared to cattle and thus allows farmers to make more efficient use of available resources.

An average dairy goat is capable of producing 2 litres of milk daily but due to lack of proper feeding and husbandry, production can drop to below 1 litre (Peacock, 1991). Shortages of water and feeds hinder the profitability of small ruminant production especially during the hot and dry seasons. In harsh environments goats produce milk when cattle have dried up. Among non-genetic factors, feeding is the main factor influencing milk composition and transformation properties (Pulina et al, 2008). An analysis of the constraints which prevents this potential from being exploited will enable recommendations to be made for project opportunities to improve the productivity of goat milk in Zimbabwe. This will ensure that smallholder farmers are able to sustain themselves through goat milk production.

1.2 Problem statement

Dairy goat production has been gaining popularity in many countries of the world in the recent years and among the small scale farmers. This has been caused by the increasing demand of goat milk due to its unique and equally important nutritional value and the fact that goats do not require large areas to keep them (Kipserem et al, 2011). However the goat sector in Zimbabwe is facing a lot of challenges.

A goat production and marketing study carried out by the Goat Forum in the six districts of Matabeleland North and South showed that farmers lose a large number of animals to mortality, instead of being used productively for consumption and sale (Homann et al, 2007). Low goat productivity keeps farmers at the stage of subsistence farming, trying to secure the survival of their existing goat herds. As a result farmers have not been realizing the full potential of goat farming as a viable source of livelihoods for a long time.

Sarupinda (2009) carried out a research on auctions opening market opportunities for smallholder farmers in western provinces of Zimbabwe. He found out that in Zimbabwe there are high levels of wastages of 45% through mortality and predators in the goat production sector. He also stated that goats in the smallholder areas of Zimbabwe have relatively low kidding rates of 76% instead of the expected value of closer to 100%. Goats are being undervalued by farmers. There is also no investment in improved husbandry, lack of business orientation and lack of formal marketing mechanisms. In Zimbabwe it was identified that high numbers of mortalities are the key challenge that restricts small-scale farmers from realizing the maximum benefits from their goats (Homann et al, 2007). It is against this background that this study sets to investigate factors influencing goat milk production in the smallholder areas of Zimbabwe.

1.3 Justification of research

Goats contribute a significant proportion towards national economies worldwide and this is projected to increase in the coming years. Small-scale farmers, especially those in remote areas, and the landless are gradually recognizing the potential of goats as a low-cost solution to their poor resource endowments. Therefore, goats deserve greater attention at both the macro and micro levels (Peacock et al, 2005).

Despite the fact that dairy goats can generate a lot of income, many smallholder farmers often do not realize these potential benefits. The increase in human population is leading to increased land pressure. This means that dairy goats are a better option since the smaller land sizes cannot support dairy cattle. If we promote goat milk production we will be able to address the millennium development goals of alleviating extreme poverty and hunger. Goat milk production in the smallholder areas of Zimbabwe is therefore a good pathway out of poverty. The research on how best we can improve goat milk production seems to be the answer in overcoming the constraints faced by smallholder farmers.

1.3.1 General objective

The general objective is to analyze the socio-economic factors that influence goat milk production in the smallholder areas of Zimbabwe.

1.3.2 Specific objectives

1 To describe goat production in the smallholder areas of Zimbabwe.

2 To determine and analyze the socio-economic factors influencing goat milk production in the smallholder areas of Zimbabwe.

3 To recommend the appropriate strategies for improving goat milk production.

1.3.3 Research hypothesis

1. Goats in the smallholder areas have not been kept primarily for milk production.

2. Cost of labor inputs per month, number of education years attained by the household head, the number of years a farmer has been rearing goats, number of does and cost of disease control per annum are the main factors influencing goat milk production.

1.3.4 Research questions

1. What are the main reasons for goat production in the smallholder areas?

2. What are the main factors influencing goat milk production?

3. What strategies can improve goat milk production?

1.4 Organization of the study

Chapter 1 has provided background information on goat milk production, research problem, justification, research objectives and research hypothesis. Chapter 2 reviews previous studies on goat milk production in Zimbabwe and from other countries. This chapter also describes the conceptual framework of the study. Chapter 3 presents the research methodology used in this study. It describes how the district, wards, villages and households were selected. It also looks at how data were processed and analyzed. Chapter 4 looks at goat production in the smallholder areas of Zimbabwe. It also describes the challenges and opportunities faced by goat farmers in the smallholder areas of Zimbabwe. Chapter 5 describes goat milk production for households in Bulilima East District. The ordinary least squares regression model is then used to analyze the socio-economic factors influencing goat milk production in the study area. Chapter 6 gives a summary of the principle findings of the study, draw conclusions and policy recommendations. The chapter ends by discussing areas for further studies.

CHAPTER 2: LITERATURE REVIEW

2.1 Introduction

The objective of this chapter is to review some of the previous studies on goat milk production. It starts with a review of the socio-economic factors influencing goat milk production. Lastly it reviews previous studies on goat milk production in Zimbabwe and other countries.

2.2 Factors influencing goat milk production

Poor nutrition is defined as one of the most serious factors limiting small ruminant productivity throughout Africa even when there appears to be an adequate quantity of vegetation available, its nutrient content may be poor (Ndlovu, 1991). This means that farmers must consider both quantitative and qualitative aspects when feeding their animals. Quantity can be increased by proper stocking of rangelands and the establishment of improved pastures to complement native pastures. Quality relates to the overall nutrient adequacy of pastures, forage and other feeds consumed as well as the measures to correct any deficiencies through improved management, fresh cut and stored forages and supplementation. Though it has been shown that the quality of feeds affect milk production in most situations smallholder farmers do not consider quality due to financial constraints and this is affects milk production (Peacock, 1996).

High yielding goats cannot meet their dietary needs with pasture grass alone and it is necessary to offer concentrate but the quantity and quality of the concentrate should be adjusted for maximum utilization of the pasture. The strategy of concentrate supplementation of grazing goats throughout the year is thus delicate (Masson et al, 1991). The inability to acquire concentrate feeds by smallholder farmers affects milk production. Moreover the level of feed intake by dairy goats is limited by roughage availabilities.

Better feed supply throughout the year may be achieved through growing species of grass with a higher nutritive value or growing legumes and fodder crops for supplementary feeding. Controlled grazing through the use of paddocks can also be used to avoid overgrazing. Continued shortage of dietary energy sources will lower milk production. Goats at an early stage of lactation need more energy.

A study on the profitability of early grass silage harvesting on dairy goats demonstrated the profitability of feeding high quality silage to goats (Flatten et al, 2012). Another study on the effects of feeding intensity during dry period on the performance of dairy goats discussed animal production factors that affect nutrition requirement as well as the necessity of forage and feed analysis for ratio balancing (Eik et al, 1991). The study defined nutrition and discussed the functions of each of the six nutrients in relation to providing that nutrient for the dairy goat. From this study the deduction that can be made is that dairy goats must be fed on all the necessary nutrients for maximum production.

Farmers must consider the quality of feeds that they give to their goats. Unfortunately, most farmers who keep goats in the tropics are not in a position to pick and choose the feeds they give to their goats according to the energy or protein content of each feed. If concentrates and forage are used at the optimum ratio, this will increase the milk's quality especially the milk fat (Van Raust et al, 2009).

Ogola et al (2010) found that feed expenses make the highest contribution to production costs. In most cases, smallholder farmers are not able to buy supplementary feeds due to lack of adequate finances and reluctance to use these inputs. The goats will not give the benefits of higher volumes of milk expected due to lack of investments in supplementary feeds by the smallholder farmers.

Onim (2009) evaluated the effectiveness of breeding and production services for dairy goat farmers and showed that poor farmers must be supported in a cost sharing manner to avoid undesirable practices. The researcher evaluated that the provision of services namely veterinary, extension, marketing, performance recording, monitoring and evaluation of dairy goat breeding activities and the provision of water in improving milk production. The effectiveness of applying these services in dairy goat farming was done and concluded that projects should take a broader responsibility of providing these services. Donkin (2000) evaluated that dairy goat cross breeds and exotics produced larger amounts of milk and sustained milk production for a larger period of 9 to 10 months. In most cases smallholder farmers in dairy goat farming are not assisted to get these services. The type and breed of goat is an important factor that affects the quantity of milk produced. Larger and heavier goats can eat more roughage than smaller ones and may be more efficient at converting poor quality food into milk than smaller ones (Steele, 1996).

Steele (1996) found out that non-dairy breeds in the tropics typically give up to 0.5 litres of milk per day. Temperate dairy goats can produce up to 5 litres of milk daily with 1-2 litres being a good average. However in most cases smallholder farmers rarely keep goats just for milk production and dairy breeds are not common. This tends to affect milk yield.

Inability to exploit the potential of a breed because the farmer is unable to fully implement management practices due to cost considerations can lead to lower benefits accruing and an enterprise being a liability than an asset (Ogola et al, 2010). In order for dairy goat farming to be successful the breeding, feeding and health care management practices must all be done correctly because a lack in any of these factors affects milk output.

Healthcare is essential to reduce production losses that arise from diseases or parasites and mortality of animals (Ogola et al, 2010). Interventions like improving nutrition or genetic improvement would be effective only if infectious diseases were curbed through preventive and curative measures (Ayalew et al, 2003). Diseases are a major constraint in dairy goat production so prevention and control is very essential. Undernourished animals are less resistant to diseases. The seriousness of diseases can be measured according to the loss in production they cause, the cost of their control or their effects if not prevented. Healthcare is essential to reduce production losses that arise from diseases or parasites and mortality of animals.

Ogola et al (2010) found out that the variations in the recommended versus the applied protocol for disease and parasite control were an indication that cost was critical for most of the farmers. The majority of smallholder farmers are not able to effectively control diseases due to high costs of veterinary drugs. Onim (1992) found out that although inputs for adoption of improved goat technologies, for example veterinary drugs, salt licks and supplementary feeds were available, prices were the major limiting factor.

The economic hardships being faced in Zimbabwe has resulted in the shortage of veterinary staff. In most cases the veterinary workers are poorly equipped with drugs and transport to attend to all the smallholder farmers. This tends to affect goat milk production. However, this does not mean that smallholder farmers must relax and wait for the veterinary staff but they must play their part in controlling diseases. Smallholder farmers can form groups so that they buy veterinary drugs as a group. Buying in groups makes drugs cheaper.

Keeping goats in a healthy condition does not necessarily mean that expensive drugs and highly trained veterinary staff must be used always. The majority of diseases affecting goats can be controlled through simple prophylactic measures. Smallholder farmers can control diseases through good feeding, vaccinations, spraying or dipping and maintaining a clean and well ventilated housing.

Homann (2007) found out that in Zimbabwe, literate farmers invested more in disease prevention, including improved housing. The study also showed that household heads with a basic level of education kept more goats especially those with secondary education. Therefore illiterate households must receive special support in order for them to improve milk production.

Cost of labor inputs influences goat milk production. Kipresem et al (2011) carried out an analysis of factors affecting dairy goat farming in Kenya. They found out that cost of labor inputs had a significant impact on the output of the dairy goat farm enterprise. Amagan and Ozden (2007) got the same results and concluded that demographic factors of farm size and labor increase the total factor productivity. This means that if farmers invest more in labor the amount of milk produced by goats will increase. Murithi (1990) found out in Kenya that if the amount of resource use is increased, then there should be substantial increase in milk production. However in most cases, smallholder farmers have few resources and as a result they do not invest in labor inputs. Some of the unnecessary goat losses through theft and predators occur as a result of inadequate labor.

The selective nature of goats during feeding and their small size make them difficult to herd even within grazing schemes that are fenced (Riviere, 1993). Goats move faster in a given grazing area and they can jump over fences into restricted areas or stray into fields. This means that there should always be someone looking after goats especially during the cropping season. This presents labor constraints. Leaving goats to graze unattended exposes them to the risk of being stolen or to be killed by predators. This means that labor inputs are very essential in goat milk production.

2.3 Review of studies from Zimbabwe

Most of the previous studies on goat production have focused on goat meat production not milk production. However these studies can be useful in trying to find the challenges faced by smallholder farmers in goat production. This is important because factors that affect goat meat production also affect goat milk production.

A Goat Forum study in 2006 on the production and marketing of goats in the six districts of Matebelaland showed that farmers lose a large number of goats through mortalities (Homann, 2007). The study also showed that goats are not being used productively for consumption and sale. Most smallholder farmers are trying to secure the survival of their existing goat herds rather than finding ways of getting into commercial goat production. Despite the fact that the Goat Forum is still young and in great need of nurturing, it has shown great potential in transforming the goat sector in the semi-arid areas of Zimbabwe.

Sikosana (2007) in his study on goat production and management found out that smallholder farmer have a major handicap in accessing information on goat management, markets, inputs and service provision. It was also observed that farmers are depending on natural rangelands for feeding their goats. Goat mortalities spike during the dry season when feed and fodder resources become scarce and are of lower nutritional quality (Sikosana, 2007).

2.4 Review of studies from other countries

Ogola et al (2010) carried out a study on dairy goat management practices in three different agro-ecological zones in Kenya. The aim of the study was to get a better understanding of the potentials and constraints of the dairy goat multiplication program that was done by Heifer Project International-Kenya (HPIK) in1994. It was observed that farmers were not following recommended feed supplements or routine disease management practices due to high costs associated with concentrates and drugs. Supplementary feeding was a major cost item. Due to cost and the perception that feeds available are sufficient to provide the necessary nutrients, 62% of the farmers did not use concentrates (Ogola et al, 2010). This study inferred that goat keeping would remain attractive for the small-scale rural producers with limited alternative ways of earning cash income.

Marketing of goats and their products is a major challenge faced by smallholder farmers. Allam (2000) found out that in France landless and resource constrained farmers sold their goats at an early age and with low market weight as they largely depend on income from them. The greatest challenge in marketing the goat milk was observed to be the competition from local cows especially during the rainy season. Although there was room for increased milk production, low and reduced milk production was related to changes occasioned by environmental and husbandry practices. Most of the farmers were unable to identify the causes of goat mortalities. The 45% goat mortality rate in the smallholder areas of Zimbabwe according to Homann (2007) statistics implies that shortages of feed and lack of investments in disease control are some of the major challenges faced in the goat production sector.

Kipresem et al (2011) carried out a study on the analysis of factors affecting dairy goat farming in Keiyo North and Keiyo South districts of Kenya. In the study it was hypothesized that farmer's factor inputs do not significantly impact on the output of the dairy goat farm enterprise. The production function used was: $Q = f(L, R, K, G, D, \mu)$... (1)

Where Q=quantity of milk produced in litres

    L= Cost of labor inputs

    R= Land under dairy goat farming

    K= Units of feeds used

    G =Number of dairy goat stock

    D =Cost of pest and disease control

    μ=error term

Kipresem et al (2011) employed a Cobb-Douglas production function to assess the impact of factor inputs on goat milk production. Cost of labor inputs, land under dairy goat farming, units of feeds used, cost of pest and disease control and number of dairy goat stock were the factor inputs.

The Cobb-Douglas production function was estimated using linear multiple regression. The explanatory power of the model was 0.94. The regression results showed that the factor inputs have a significant effect on goat milk production. The sum of elasticities was 1.946 showing that doubling the inputs will more than double the amount of goat milk produced.

Table 1: Independent Variables Used in the Study

Independent variables	Method of measuring used
Land under dairy goat farming	The total area of land in hectares owned by each farmer was recorded.
Cost of labor inputs	Labor number and total cost of the wages per month was recorded.
Number of dairy goat stock	The total number of does owned by a farmer was recorded.
Cost of pest and disease control	Quantities and total cost of all veterinary drugs and equipment was recorded.
Units of feeds	Quantity of fodder feeds given to a goat that is quantity of feed concentrates, mineral salts, and other feeds like banana leaves, sacks of natural pasture or shrubs, acacia pods in kilograms.

Kipresem et al (2011) concluded that cost of labor inputs, land under dairy goat farming, and units of feeds used, number of dairy goat stock and cost of pest and disease control affect dairy goat farming.

Kipresem et al (2012) undertook a study for analyzing the productive allocative efficiency of factors of production used in dairy farming. The study was done in 2011 in Elgeyo-Marakwet area in Kenya. The elasticities of factor inputs were obtained from a linear multiple regression analysis. Labor hours, feeds in kg and frequency of disease and pest control were the factor inputs for milk production. Results of this study indicated that the use of available feeds as well as cost of disease control and labor inputs were not used efficiently implying that there was no optimal allocation of these resources. This affected goat milk production.

Shah et al (2009) made an attempt to measure the net profit scenarios for goats and their products and to estimate their economics of production in Pakistan. Results of the study indicated that goat milk production is a function of yield of milk per day, lactation period and the characteristics, breed and age of animal reared. The kind and quantity of concentrate fed to the animal also influences goat milk yield. Shah et al (2009) used the Ordinary Least Squares regression method. The model was specified as follows:

$$Y_i = \beta_0 + \beta_1 FS + \beta_2 MA + \beta_3 FD + \beta_4 L + \mu \dots\dots\dots\dots\dots\dots\dots\dots\dots\dots\dots (2)$$

Where Y= income from milk production

      FS=farm size

      MA=milking animal

      FD=feed cost (green and dry fodder)

      L=labor cost

      β_0=intercept

      β_s =coefficient with respect to FS, MA, FD and L.

The explanatory power of the model was 0.641. Shah et al (2009) also used a Cobb-Douglas type of production function in the same study to trace out the scale of returns for goat milk production. The Cobb-Douglas production function was as follows:

$$Y = C(FS)^{\beta_1} (MA)^{\beta_2} (FD)^{\beta_3} (L)^{\beta_4} \dots\dots\dots\dots\dots\dots\dots\dots\dots\dots\dots..(3)$$

Where Y, FS, FD, MA and L are as defined earlier and carry the same meaning and C is a constant. The Cobb-Douglas production function was then estimated using the ordinary least squares regression method after transforming it into a linear function by expressing it in logarithm form as follows:

$$\log Y = \log C + \beta_1 \log FS + \beta_2 \log MA + \beta_3 \log FD + \beta_4 \log L + \mu \dots\dots\dots\dots\dots\dots\dots(4)$$

Results of the regression showed that the sum of elasticities was 1.28. This means that doubling the inputs will more than double the amount of goat milk produced.

Torane (2009) while analyzing the effect of green fodder concentrates, veterinary expenditure, and human labor on the income of a goat enterprise in the North Konkan region of Maharashtra used a Cobb-Douglas production function. The Cobb-Douglas production function was specified as follows:

$$Y4 = a.X_1^{b1}.X_2^{b2}.X_3^{b4}.X_4^{b4}.X_5^{b5} \dots\dots\dots\dots\dots\dots\dots\dots\dots\dots\dots\dots\dots (5)$$

Where Y4=income (Rs. / animal)

X_1=green fodder (Rs. /animal)

X_2=concentrates (Rs./animal)

X_3=veterinary expenses (Rs. /animal)

X_4=human labor (Rs. /animal)

a=intercept

$b_1, b_2, b_3 \dots\dots\dots b_n$= regression coefficients.

A multiple regression was done and it was observed that all the variables had a significant effect on income from goat production.

The literature review under so far indicates that the majority of researchers used an Ordinary Least Squares regression method. In most of these previous studies, a Cobb-Douglas production function was also estimated using the ordinary least squares regression method. The idea behind the use of the Cobb-Douglas production was for the calculation of the scale of returns on goat milk production.

Among the various socio-economic factors influencing goat milk production, labor cost, veterinary expenses, feed cost, farm size, quantity of concentrates, number of dairy stock, were excessively used in most of the studies. However many of these researchers did not include the effect of farming experience on goat milk production. For the socio-economic factors influencing goat milk production in the smallholder areas of Zimbabwe this variable will be included.

However for the socio-economic factors influencing goat milk production in the smallholder areas of Zimbabwe, some of these factors are not applicable despite the fact that they were found to have an impact on goat milk production. Land under dairy goat farming and units of feeds used are the variables that cannot be assessed to see their impacts on goat milk production in Zimbabwe. In the Zimbabwean situation there is no private ownership of grazing land hence it is difficult to quantify the area of land under dairy goat farming for each farmer because the land is shared among all the farmers in the community. This limits the area under which a farmer can grow forages to feed his goats. Ninety-nine percent of goats in Zimbabwe are managed under the traditional husbandry system in which they are wholly dependent on rangelands for nutrition with no supplementary feeds during the dry seasons (Sibanda, 1993). This means it is also difficult to assess the effect of units of feeds on goat milk production because goats are kept under a free range grazing system hence it is difficult to quantify the amount of feed grazed by each goat.

2.5 Econometric model specification for goat milk production in the smallholder areas of Zimbabwe

The ordinary least squares has been considered to be the efficient way of regressing socio-economic factors influencing goat milk production against milk output for smallholder farmers in Zimbabwe. The econometric model is specified as follows:

$$Y_i = \beta_0 + \beta_1 X_1 + \beta_2 X_2 + \beta_3 X_3 + \beta_4 X_4 + \beta_5 X_5 + \mu \dots \dots \dots (6)$$

Where Y_i=volume of goat milk in litres produced by a goat per day.

β_0=constant.

$\beta_1 - \beta_5$= beta coefficients

X_1=number of education years attained by the household head.

X_2=number of years a farmer has been rearing goats.

X_3= cost of disease control per annum.

X_4= cost of labor inputs per month.

X_5=number of does

μ=error term

2.6 Conceptual Framework

The conceptual framework adopted in this study is built on the relationships between household characteristics, non-household characteristics and goat milk production. The framework will highlight all the possible links between these various aspects.

Figure 2.1: Conceptual Framework for goat milk production

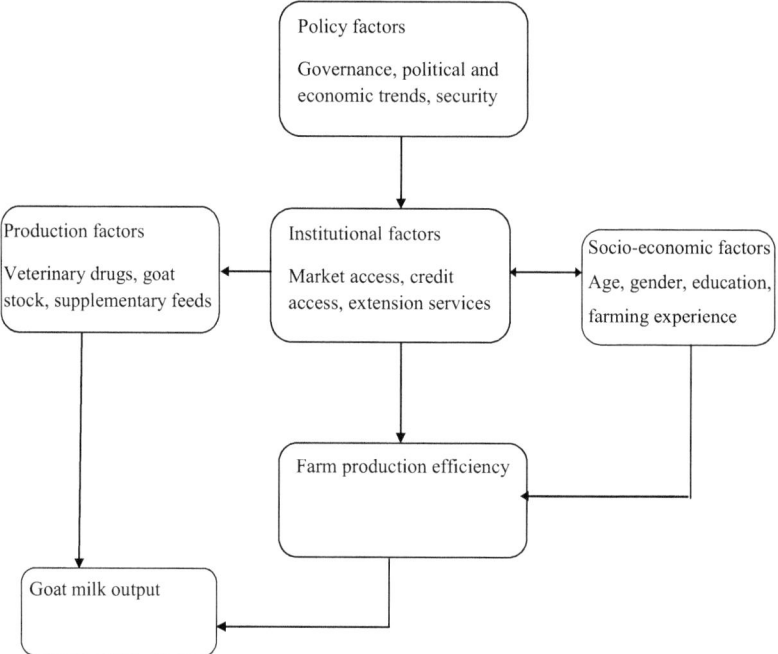

The policy environment has an influence on goat production. It affects the access and provision of credit and extension services to goat farmers. The production factors such as supplementary feeds, veterinary drugs, supplies and goat stock are used as inputs in the production of goat milk.

The availability and distribution of these inputs is affected by the policy framework in place and this determines the level of goat milk production.

Institutional factors such as credit access and extension services affect goat milk production. Credit access enables farmers to acquire inputs for goat milk production. Extension services provide farmers with information on better methods of goat milk production.

Socio-economic factors such as gender, age, number of household members, farming experience and level of education also affects goat milk production. The manner in which the available resources are used by the farmer is crucial because it affects how all the inputs are changed into milk output.

2.7 Summary

The literature reviewed in this chapter has explained that cost of labor inputs, cost of disease control, quantity of feeds, number of dairy goat stock affect goat milk production. From these socio-economic factors, land under dairy goat farming and quantity of feeds cannot be included in the study of factors influencing goat milk production in Zimbabwe. In the smallholder areas of Zimbabwe, grazing lands are communally owned making it difficult to measure land under dairy farming or quantity of feeds for each farmer. Most of the previous studies on goat milk production did not include the variable of goat farming experience among the variables affecting on goat milk production. This variable will be assessed to find its impact on goat milk production in the smallholder areas of Zimbabwe.

CHAPTER 3: RESEARCH METHODS

3.1 Introduction

This chapter addresses the major methodological procedures that were followed in carrying out the study; issues related to the choice of study area, research design, research instruments and sampling procedures.

3.2 Study Area

The study was conducted in Bulilima East District which lies in Matabeleland South Province and was purposively selected because milking of goats is a common practice. Two wards in the district were purposively selected for study on the basis of information provided by the local extension office of the Department of Agricultural Technical and Extension Services (Agritex). One ward where the majority of farmers milked their goats (Ward 21) and another (Ward 2) where the majority of farmers do not milk their goats were selected.

Using a list of villages in each ward obtained from Agritex one village from each ward was randomly selected. The villages finally selected for the study were Sibomvu village in Ward 21 and Gwambe village in Ward 2. The research targeted two villages only due to limited research resources available for the study. A list of goat owners in the two villages was obtained from the village head. Using simple random sampling, twenty households were selected from each village, giving a total of 40 households.

3.3 Data collection

Data were collected through interviews with the selected households using a questionnaire developed for the study. The questionnaire was designed to collect broad categories of data on (1) demographic characteristics of households (2) ownership and functions of goats (3) milk output and socioeconomic production constraints (4) labor management and cost of inputs and (5) health care management practices, costs and mortality.

3.4 Data analysis

Data were analyzed using the Statistical Package for Social Sciences (SPSS) version 16. Descriptive statistics were used to summarize socioeconomic indicators. Multiple ordinary least squares regression was performed to assess factors influencing goat milk production. The econometric model used was

$$Y_i = \beta_0 + \beta_1 X_1 + \beta_2 X_2 + \beta_3 X_3 + \beta_4 X_{4+} + \beta_5 X_{5+} \mu$$

Where Y_i = volume of goat milk in litres produced by a goat per day.

β_0 = constant.

β_1 - β_5 = beta coefficients.

X_1 = number of education years attained by the household head.

X_2 = number of years a farmer has been rearing goats.

X_3 = cost of disease control per annum.

X_4 = cost of labor inputs per month.

X_5 = number of does.

μ = error term.

3.5 Limitations of the study

The majority of the respondents did not keep goat farming records so they relied on memory to recall. This reduced the quality of data that was collected. Due to financial constraints the sample size used was not big enough to detect significant differences in some cases.

Table 3.1 gives a summary of the analytical tools that were used in testing the research hypothesis.

Table 3.1: Summary on Analytical tools and hypothesis testing

Objectives	Research hypothesis	Data required -Variables	Type of Analysis
1. To describe goat production in the smallholder areas of Zimbabwe.	Goats in the smallholder areas have not been kept primarily for milk production.	Functions of goats, Breeds, goat statistics, outflows of goat herds, marketing	descriptive statistics trend analysis Graphical analysis. T-test
2. To determine the main factors influencing goat milk production	Cost of labor inputs, number of education years a farmer attained by household head, number of years a farmer has been rearing goats, number of does and cost of disease control per annum are the main factors influencing goat milk production.	volume of milk in liters produced per day, number of hired labor and cost per month, highest level of education attained by household head, number of years a farmer has been keeping goats, types of veterinary drugs and supplies and cost per year, transport costs Number of does owned by a farmer.	Ordinary least squares multiple regression analysis
3.To recommend strategies that will improve goat milk production			

Table 3.2 gives a summary of the apriori expectations of the variables in the econometric model to be tested.

Table 3.2: Socio-economic factors influencing goat milk production and their hypothesized effects

Explanatory variables	Hypothesized effect	Explanation of the variable effect
Cost of labor inputs	Positive	If farmers invest more in labor inputs level of goat milk production will increase.
Number of education years attained by the household head.	Positive	Education increases farmer's ability to get and use information. This variable is therefore assumed to have a positive relationship with the volume of milk produced.
Number of years a farmer has been rearing goats	Positive	A household with more farming experience will be more productive so there is a positive relationship.
Number of does	Positive	A farmer with many does has more goat farming experience.
Cost of disease control per month	Positive	If farmers invest more in disease control his goats will become healthier and more productive.

CHAPTER 4: GOAT PRODUCTION IN THE SMALLHOLDER AREAS OF ZIMBABWE

4.1 Introduction

The objective of this chapter is to describe goat production in the smallholder areas of Zimbabwe. This will give insights on challenges and opportunities faced by goat farmers and their impacts on goat milk production.

4.2 Goat populations in Zimbabwe

The majority of goats in Zimbabwe (40.7%) are found in Matabelaland Provinces since they can survive under harsh conditions. Mashonaland West Province has the least number of goats (2.8%) because of lower temperatures which are not suitable for goat production (Table 4.1).

Table 4.1: National estimates and distribution of goats in Zimbabwe by province in 2012

Province	Goat population	Share %
Manicaland	634742	20.7
Mashonaland Central	227751	7.4
Mashonaland East	154502	5.0
Mashonaland West	85546	2.8
Matabelaland North	805884	26.2
Matabelaland South	445827	14.5
Masvingo	289180	9.4
Midlands	429418	14.0
Total	3072850	100.0

Source: Ministry of Agriculture, Mechanisation and Irrigation Development (2013)

4.3 Trends in goat population in Zimbabwe (2003-2012)

There is a constant trend in goat population from 2003-2007(Figure 4.1). There was a slight decline in goat population from 2007 up to 2009 as a result of the economic hardships that were being faced in the country. Most farmers were selling and slaughtering their goats to survive leading to a decline in goat numbers. Since dollarisation of the economy in 2009, the goat population is increasing because farmers are getting more incentives to produce goats. They are investing in goat production because they are able to get a market for their goats thereby improving their level of production.

Figure 4.1 Goat population trends in Zimbabwe (2003-2012)

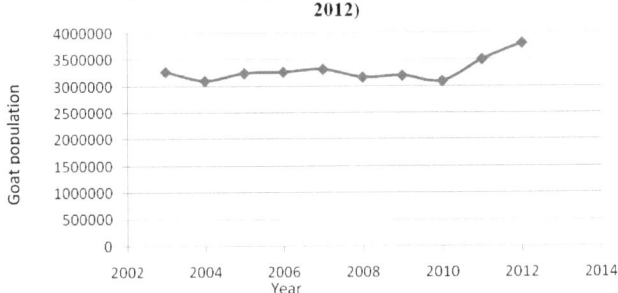

Source: Ministry of Agriculture, Mechanisation and Irrigation Development (2013)

4.4 Goat production in the smallholder areas of Zimbabwe

The changes of land tenure policies that occurred in Zimbabwe in 2000 caused a decline of the country's agricultural capacity. This was further worsened by the harsh economic environment which was characterised by shortages and price increases of farming inputs. The reduction of commercial cattle herd (-75% from 1996 to 2004) led to higher beef price increases and stimulated consumers to subtitute it with goat meat (Sibanda, 2005). In this respect goats are offering a good alternative.

This has created an important opportunity for the goat industry because goat meat can serve as a ready alternative to beef given that the population of goats in Zimbabwe is currently at 3 million (FAO, 2005). Goats rank second to cattle in terms of economic importance in the livestock sector of Zimbabwe. Goat numbers have been steadily increasing and there is a greater scope for goat market development in the urban and rural areas (Sibanda, 2005).

Zimbabwe's small stock comprises of 3.3 million goats and 390000 sheep. Approximately 97% of the goats are owned by smallholder farmers (van Rooyen et al, 2007). In Zimbabwe goat production is a nationwide activity although the semi-arid regions are the most productive areas. The semi-arid areas of Zimbabwe are found in Matebeleland Provinces and 40.7% of the total goats in Zimbabwe are found in these semi-arid areas. Goats can survive under harsh conditions compaired to cattle and it is for this reason that they are found in large numbers in the semi-arid areas of Zimbabwe. Goat milk production is undertaken in the drier ecological zones of natural region 4 and 5. However smallholder farmers do not keep goats primarily for milk production but milk is one of the products of simply keeping goats. The goat milk production is mainly for home consumption and kid feeding. It is not an organised and controlled production sector as in the case for cow milk.

4.4.1 Goat breeds

The majority of goat breeds in Zimbabwe are the indigenous goats. These are the Mashona and Matebele goat which are for meat production. The other breeds are the Boer, Angora and Saanen goats (Table 4.2).

Table 4.2: Main breeds in Zimbabwe

Breed	Mature mass female (kg)	Products
Indigenous	25-40	Meat, skin
Boer	65	Meat, skin
Saanen	65	Milk
Angora	25-30	Mohair

Source: Chimowa (2008)

In Zimbabwe, the local goat breeds are mainly found in the communal areas whilst the exotic breeds are found in the commercial areas (Siringwani et al, 2010). The indigenous Matebele and Mashona breeds are the predominant type of goats in the communal land areas (Chifamba et al, 1993).

Uncontrolled breeding is a common practice in the communal areas of Zimbabwe. Therefore smallholder farmers are failing to get the benefits associated with improved breeding. Farmers do not plan the mating periods of their goats so goats tend to kid at any time of year regardless of whether feed is available or not. This explains why kid mortality rates in Zimbabwe are very high.

4.4.2 Importance of goats in the smallholder areas of Zimbabwe

A study on goat production and marketing commissioned by the Goat Forum in 2006 revealed that farmers value the multiple functions of goats. Farmers ranked income as the most important followed by meat, milk, manure and traditional rituals (Homann et al, 2007). In this study it was also found out that Zimbabwean goats specifically, are utilized to supplement household food requirements and sold to purchase food items and fund educational expenses.

Goats complement the value of cattle and their role is increasing in the rural areas of Zimbabwe (van Rooyen, 2009). They also contribute significantly to the educational sector by generating funds to cater for educational expenses. This is a very important contribution to the livelihoods of the smallholder farmers in the sense that if they get education, their ability to manage livestock becomes more efficient. Livestock production will then increase thereby improving their standards of living.

Van Rooyen (2009) carried out a study on the importance of goats in the semi-arid areas of Zimbabwe. It was shown that the role of goats to the livelihoods of the small scale farmers is increasing. Most smallholder farmers are relying on goats as a source of livehood. Since goats are not a source of draft power most farmers rely on them for meat more than they rely on beef.

Goats also contribute to income more than cattle because goats are easy to convert to cash compared to cattle.The high profitability and fast turnover of goats due to their earlier maturity and shorter generation interval makes them a more reliable source of income as compared to cattle. Goats' smaller size and rapid growth rates makes them a flexible short term form of investment than cattle. Where gardens are functional, goat manure is used to improve soil structure and fertility in the smallholder areas of Zimbabwe. Figure 4.2 shows the functions and importance of goats in the semi-arid areas of Zimbabwe.

Figure 4.2: The functions and relative importance of goats and cattle in semi-arid Zimbabwe

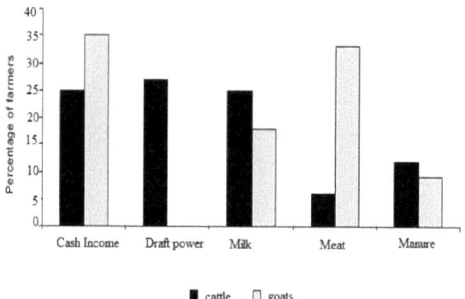

Source: van Rooyen (2009)

4.4.3 Goat Ownership Patterns

According to Homann (2007) statistics, the majority of households in Zimbabwe (52%) have less than 8 goats, 30% have between 8 and 20 goats and only 18% have more than 20 goats. Since smallholder farmers own very small herd sizes it means that they cannot afford to lose many goats through mortalities.

4.5 Challenges in goat production

4.5.1 Goat feeding practices

Ninety-nine percent of goats in Zimbabwe are managed under the traditional husbandry system in which they are wholly dependent on rangelands for nutrition with no supplementary feeds during the dry seasons (Sibanda, 1993). Goats in the smallholder areas are kept under a free range grazing system in which they are kept in pens overnight and share communal grazing lands during the day.

Although farmers rely on natural pasture which is freely available, many other supplements which are rather costly are necessary for balancing the diet to meet the nutrient requirements of goats. During the dry season, goat production is greatly affected by the inadequate availability and poor quality of feed resources. Most goats die during these dry seasons since the majority of smallholder farmers do not have supplementary feeds. Shortages of feeds make goats in the smallholder areas more susceptible to diseases. However during the wet seasons, rangelands are able to support the nutritional requirements of goats.

Homann (2006) found out that the majority of farmers (93%) in Matebelaland face shortages in the supply of sufficient feed for optimal goat production. Homann (2005) carried out a study on the importance and development challenge on goats in Beitbridge, Binga, Gwanda, Matobo, Nkayi and Tsholotsho districts. In all the districts it was noticed that farmers face feed shortages from July up-to December. During this period feeds deteriorated in both quality and quantity. The supply of rangeland resources improves during the rainy season and gets depleted on the onset of the cold dry season.

4.5.2 Goat mortality rates

Kid mortality is the major cause of losses in goat production. Matopos Research Station recorded an average pre-weaning kid mortality of 17.4% over a period of ten years (Chimowa, 2008).

Goat mortality is one of the greatest challenges faced by smallholder farmers in Zimbabwe. Since most of the smallholder farmers rely on natural rangelands, the cold dry season is a period of feed shortages. This results in very high goat mortalities during the dry seasons. It causes very high losses especially among kids and lactating females.

Homann (2007) carried out a study on promoting goat markets and technology development in semi-arid Zimbabwe. He observed that 45% of goats are lost through mortalities yet only 24% were sold and 14% slaughtered for home consumption (Figure 4.3). From these findings we can deduce that goat losses due to deaths greatly exceed the profitable use of goats. Smallholder goat farmers are not realizing the potential benefits of goats. The small ruminant's potential is being hindered by factors such as health problems, poor nutrition, breeding, marketing and predator losses. Goat mortalities are very high during the dry seasons due to seasonal changes in the quality and availability of feed resources. Shortages of feed cause the goats to be more susceptible to diseases. Smallholder farmers do not have the resources to make investments that will ensure that goats will survive during dry seasons. They continue to lose a large number of animals through mortalities (Homann, 2007). Figure 4.3 shows the monthly outflows in goat herds in Zimbabwe from 2005 to 2006.

Figure 4.3: Monthly outflows in goat herds in Zimbabwe.

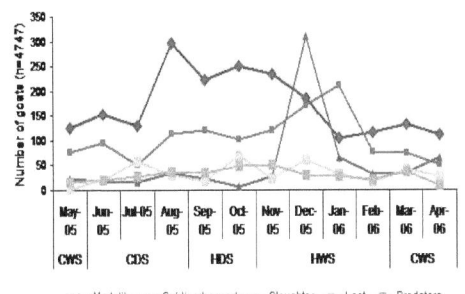

Source: Goat reconnaissance survey by van Rooyen and Homann-Kee Tu (2007).

4.6 Goat marketing

Goat marketing in Zimbabwe is mainly through the informal sector. According to Chimowa (2008) goats are marketed through the following outlets:

i) Informal markets

ii) Private abattoirs and butcheries

iii) Community abattoirs

iv) Private slaughtering

v) Cottage industries for mohair and skins

vi) Cold Storage Company.

Direct sales from producer to local consumer are probably the most important way of marketing goats in the smallholder areas of Zimbabwe. The commercial market for goats in the country is underdeveloped. Weak public and private sectors are failing to provide the necessary inputs and know-how (Hargreaves et al, 2005). The input and output markets for goats are not functioning well because of poor infrastructure in the country especially in the smallholder areas. Inputs such as veterinary drugs and commercial feeds are not accessible to the smallholder farmers because of very high procurement costs.

On the output market buyers and goat processors are getting low quality goat meat and the supply is inconsistent. Despite the great potential, the market demand for goat meat cannot be met by the current supply volumes and on average there is a demand gap for goat meat of at least 60% (Nyathi et al, 2007).

Farm gate sales is the major market channel although it disadvantages famers because there is no grading system, competition is low and farmers have low bargaining power. Market prices from the farm gate sales are very low. At farm gate sales, although farmers incur low costs, they are in an unfavorable position to negotiate prices, especially when they are in dire need for cash. The other informal market channel common in the smallholder areas of Zimbabwe is the barter exchange system. Social marketing of goats for example gifts, lobola and exchanges are very common in the smallholder areas.

According to a goat production baseline survey by the Goat Forum in 2006, smallholder farmers in Matebelaland sold their goats to local traders at very low prices. Despite the fact that farmers owned large numbers of goats they could not earn enough income. The goat marketing system is inefficient in the smallholder areas. To rectify the situation the Netherlands Development Organization proposed the goat auction system in 2008.

4.6.1 Goat auction system in Zimbabwe

Netherlands Development Organization introduced the goat auction system in Zimbabwe in 2008. The auctions were meant to increase transparency of transactions, reduce transaction costs for the buyers, and promote supply and demand mechanisms through bidding processes and to increase access to market information. Technical support of the Goat Auctions comes from Department of Agricultural Technical Extension Services, Organization of Rural Association for Progress and Netherlands Development Organization.

The goat auction system was a good approach to improve the marketing system of goats because the huge numbers of goats in the drier regions of the country could attract many buyers. The first goat auctions were held in Beitbridge and Gwanda in 2008. Goat auctions greatly improved the marketing system of goats in the country. More than 1500 farmers have since benefited from the pilot auctions, earning prices which are more than three times higher than what they used to get at the farm gate level (Sarupinda, 2009).

The goat auction system has improved the marketing of goats in the smallholder areas. It increased farmer's access to market information and they were exposed to more buyers. Sale pens were put in place and they are being utilized for the auctioning of goats. There is in an improvement in grading system of goats.

4.7 Summary

Goats in the smallholder areas of Zimbabwe play a significant role in the livelihoods of people directly through meat and milk production and indirectly by increasing their cash income. Despite the fact that the goat sector has a potential of improving livelihoods in rural Zimbabwe, the goat sector faces a lot of challenges. Inadequate nutrition, prevalence of diseases and lack of economic incentives are some of the major challenges. This means that strategic plans must be put in place so as to strengthen the potential of the goat sector and help the smallholder farmers to get the maximum from their goats.

CHAPTER 5: GOAT MILK PRODUCTION IN SMALLHOLDER AREAS OF ZIMBABWE

5.1 Introduction

This chapter describes goat production in the smallholder areas of Zimbabwe. The chapter starts with descriptive statistics for the households in Bulilima East District. Goat milk production is then described and the hypothesis that goats in the smallholder areas have not been kept primarily for milk production is then tested. This chapter ends by addressing the hypothesis that cost of labor inputs per month, number of education years attained by the household head, the number of years a farmer has been rearing goats, number of does and cost of disease control per annum are the main factors influencing goat milk production.

5.2 Demographic characteristics of household heads in Bulilima East District, Zimbabwe 2013

The first objective of the study was to describe goat production in the smallholder areas of Zimbabwe. This section describes goat production in the smallholder areas of Zimbabwe. Table 5.1 shows the descriptive statistics of the household heads rearing goats in Bulilima East District.

The average age of the household heads was 44 years (Table 5.1). The average age of household heads milking goats is higher than that of farmers who do not milk their goats. The majority of household heads (90%) have received formal education (Table 5.1) which suggests that communication of technical knowledge on goat milk production will be easy. Farmers who milk their goats are more educated than those not milking goats (Table 5.1). None of the household heads has an agricultural qualification.

Table 5.1: Demographic characteristics of household heads in Bulilima East District, Zimbabwe 2013

	Households milking goats n=20	Households not milking goats n=20	Total N=40
Household head: % Male	65	50	57.5
% Female	35	50	42.5
Mean age of household head	48(15.02)	41(13.22)	44(14.47)
Mean number of household members	7(2)	6(2)	7(2)
Education level of household head : % None	15	5	10
% Primary	50	45	47,5
% Secondary	25	45	35
% Tertiary	10	5	7.5
% with any Agricultural qualification	0	0	0
Mean number of years in goat farming	18.10(13.56)	18.05(14.91)	18.08(14.07)

Source: Survey data (2013).

Note: The numbers in brackets are the standard deviations

Table 5.2 shows the descriptive statistics for the factors related to goat milk production in Bulilima East District. The household heads have an average of 10 goats distributed as follows; bucks 20% of the herd, does (60%) and kids (2%) (Table 5.2). The average volume of milk produced by a doe per day was 0.29 litres. The cost of disease control per annum is $18 per annum whilst labor inputs costs $ 35 per month.

Table 5.2: Characteristics of factors related to goat milk production for the household heads in Bulilima East District, Zimbabwe 2013

Variable	Mean(SD)	Minimum	Maximum
Total number of goats	10(7)	2	32
Number of bucks	2(2)	0	6
Number of does	6(4)	1	18
Volume of goat milk in litres per day	0.29(0.29)	0	0.70
Number of kids	2(2)	0	7
Average kid mortality	3.25(2.46)	0	10
Cost of disease control per annum	17.97(23.10)	0	71
Cost of labor inputs per month	35.31(9.36)	10	60

Source: Survey data (2013)

5.3 Comparisons of characteristics between farmers milking goats and those not milking, Bulilima East District, 2013

Figure 5.1 shows the cost of disease control and labor inputs for farmers milking goats and those not milking goats. The average cost of labor inputs per month for farmers who milk their goats was $35.53 and $ 34.50 for farmers not milking goats. Farmers who do not milk their goats had the highest cost of disease control per annum with an average of $19.85 compared to $16.05 for those milking their goats.

Figure 5.1: Cost of labor inputs and disease control for farmers who milk goats and those who do not milk their goats, Bulilima East District, 2013

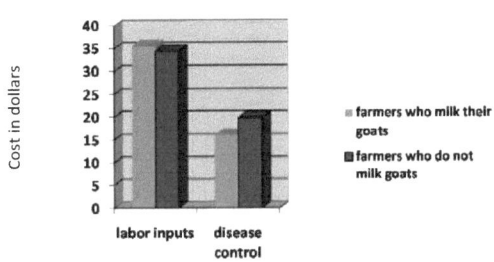

Source: Survey data (2013)

In order to assess whether the cost of labor inputs and cost of disease control in the two groups of farmers was significantly different, a T-test was performed (Table 5.3).

Table 5.3: Independent sample t-test for cost of labor inputs and cost of disease control

	Farmers milking goats	Farmers not milking goats	T	Sig
Cost of labor inputs	35.5	34.5	0.342	0.734
Cost of disease control	16.05	19.85	-0.515	0.609

Source: Survey data (2012).

The results indicate that the difference between the cost of labor inputs for farmers milking goats and those not milking is not significant. The probability of the t-statistic (0.342) for the cost of labor inputs is 0.734 which is not significant at 5% level of significance. This result can be attributed to the fact that labor units required for the two groups of farmers are almost the same.

Besides the process of milking, there is no other goat farming activity done by farmers who milk their goats in addition to the normal activities done by farmers who do not milk goats hence labor costs are not significantly different.

Another observation made is that the difference between the cost disease control between farmers milking goats and those not milking is not significant. The probability of the t-statistic (-0.515) for the cost of disease control is 0.609 which is not significant at 5% level of significance.

5.4 The main reasons for goat production in the smallholder areas of Zimbabwe

The first hypothesis of the study was that goats in the smallholder areas of Zimbabwe have not been kept primarily for milk production. In order to test this hypothesis, farmers were asked the main reasons for keeping goats. The descriptive statistics show that goats in the smallholder areas are not kept primarily for milk production but milk is simply a product of keeping goats. The main reason for goat farming in the study area is income generation for 72.5% of the households and only 5% of the farmers keep goats primarily for milk production (Figure 5.2). This means that we can conclude that goats in the smallholder areas are not kept primarily for milk production.

Figure 5.2: The main reasons for goat production in the smallholder areas of Zimbabwe.

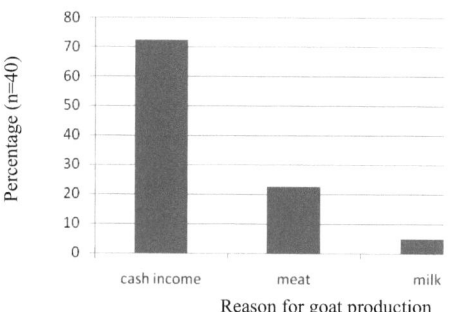

Source: survey data (2013).

In order to assess the main reasons for not milking goats, those farmers who do not milk their goats were asked their main reasons for not milking goats (Figure 5.3).

Figure 5.3 Reasons for not milking goats

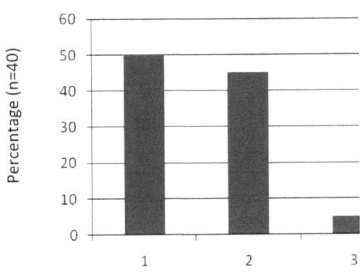

1- Goat milk is not consumed

2 -Little quantities produced

3-Cultural reasons

Source: Survey data (2013)

The results show that the majority of smallholder farmers (50%) do not milk their goats because they do not consume the milk. Informal discussions held with farmers indicated that the main reason farmers do not consume goat milk was due to tastes and preferences. The other reason for 45% of the households not milking goats is that their goats produce very little quantities of milk. Only a few farmers (5%) do not milk their goats because of cultural reasons.

5.5 An econometric analysis of the socio-economic factors influencing goat milk production in the smallholder areas of Zimbabwe

The second objective of the study was to determine the main factors influencing goat milk production. This section will analyze these factors and also test the hypothesis that cost of labor inputs per month, number of education years attained by the household head, the number of years a farmer has been rearing goats, number of does and cost of disease control per annum are the main factors influencing goat milk production. The hypothesis will be tested using multiple regression analysis.

5.5.1 Regression analysis of the socio-economic factors influencing goat milk production in the smallholder areas of Zimbabwe

A multiple regression model with 5 explanatory variables was used to analyze the relation of the total goat milk output in response to changes in explanatory variables. Cost of labor inputs per month, cost of disease control per annum, number of years a farmer has been rearing goats, number of does and number of education years attained were the explanatory variables. Volume of goat milk produced per day in litres was the dependent variable.

5.5.2 Regression Model

The regression equation $Y_i=\beta_0+\beta_1X_1+\beta_2X_2+\beta_3X_3+\beta_4X_4+\beta_5X_5+\mu$ for goat milk production in the smallholder areas of Zimbabwe was estimated using the backward method of the ordinary least squares regression technique. The backward method was used because it drops all the insignificant variables until a good model with significant variables is obtained. In this regression analysis five models were obtained using the backward method but model 2 was chosen because it had the largest adjusted R^2_{adj} (0.606) (Table 5.4).

Table 5.4: Model summary

Model	R	R Square	Adjusted R^2	Durbin Watson
1	.812	.659	.590	
2	.812	.659	.606	
3	.800	.639	.599	
4	.790	.623	.597	
5	.767	.589	.575	2.228

Source: Regression of survey data (2013)

Table 5.5 shows the results of the regression of the socio-economic factors influencing goat milk production in the smallholder areas of Zimbabwe.

Table 5.5: Results of Regression Analysis

Variable	Beta coefficient	T-statistic	Significance
X_1=number of education years attained by household head	0.14	2.075	0.048**
X_3=cost of disease control per annum	0.006	5.208	0.000**
X_4=cost of labor inputs per month	0.04	1.215	0.235
X_5=number of does	-0.09	-1.346	0.190
R=0.812 R^2=0.659 Adjusted R^2=0.606 Durbin Watson=2.228 β_0=-0.018			

**Significant at 5%.

Source: Survey data (2013)

From the regression analysis, variables X_4 and X_5 were found be insignificant and X_2 was excluded from model 2. Therefore the final estimated model is:

$$Y_i = \beta_0 + \beta_1 X_1 + \beta_3 X_3 + \mu \dots\dots\dots\dots\dots\dots\dots\dots\dots \text{(3)}$$

Substituting the values of β_0, β_1 and β_3 into equation 3 the regression model becomes:

$$Y_i = -0.018 + 0.14 X_1 + 0.006 X_3 + \mu \dots\dots\dots\dots\dots\dots\dots\dots\dots\dots\dots\dots\dots\dots\dots\dots \text{(4)}$$

5.6 Assessment of the strength of the econometric model

In order to assess the strength of the model, a number of tests were carried out. These were goodness of fit, and tests for the presence of autocorrelation and multicollinearity. The results are reported in the following sections.

5.6.1 Measure of goodness of fit (coefficient of determination)

Adjusted R^2 is the proportion of total variation in goat milk production accounted for by the predictors in the linear regression model. Predictors in the model have been able to account for 60.6% variability in the goat milk productivity. It shows that the data is best described using the linear model since it exhibits more linear relation between the independent variables and milk productivity. Adjusted R^2 for the relationship between the set of independent variables and the dependent variable is 0.606, which would be characterized as strong because it is closer to 1.

5.6.2 Test for the presence of autocorrelation

The Durbin-Watson test (DW) was used for testing for the presence of autocorrelation. As a rule of the thump the value of DW should fall within the range of 1.5 to 2.5 to suggest the absence of autocorrelation. From model 2 used in regressing factors influencing goat milk production, the Durbin-Watson value is 2.228 which implies the absence of autocorrelation since it lies in the critical region.

5.6.3 Test for the presence of multicollinearity

In order to assess that there were no perfect linear relationships among the explanatory variables in the model, a test for multicollinearity was carried out. The results of the test are reported in the correlation matrix table (Table 5.6).

5.6.4 Correlation matrix for socio-economic factors influencing goat milk production in the smallholder areas of Zimbabwe

The correlation matrix was used to assess the presence of multicollinearity. A correlation greater than 0.7 would mean that the independent variable factors are highly correlated and one of the factors will be dropped from the model to avoid spurious results. The correlation matrix is shown in Table 5.6.

Table 5.6: Coefficient Correlation Matrix Table

	X_1	X_3	X_4	X_5
X_1	1.000			
X_3	-0.387	1.000		
X_4	0.166	-0.188	1.000	
X_5	-0.356	0.054	-0.253	1.000

Source: Survey data (2013)

The correlation matrix results show that none of the variables has a higher correlation to cause effects associated with the presence of multicollinearity. This is because none of the correlations is greater than 0.7. Therefore all the variables are considered. Multicollinearity is not a problem in this regression analysis.

5.7 Discussion of Results

The study shows that despite the potential, goats in the smallholder areas of Zimbabwe are not kept primarily for milk production. The main reason for goat production is for income generation (about 73% of the households). This is mainly because the study area lies in natural agro-ecological region V which receives less than 450 of rainfall per annum. The region is not suitable for crop production and livestock production is the predominant type of enterprise. The region also carries 55% of the estimated three million goat population in Zimbabwe (MAMID 2012).Van Rooyen (2009) in a study on the importance of goats in the semi-arid areas of Zimbabwe also found out that the main importance for goat production is income generation. This implies that the current scenario for goat production does not favor adoption and easy sustainability of goat milk production in the smallholder areas of Zimbabwe. Results show that tastes and preference are an important determinant in the households' decision to milk goats. This is in line with studies done by Boor et al (1987) in Kenya, Manyenga (1987), Mowlem (2005) in UK and Yangilar (2013) which indicated that the strong flavor and taste limited the consumption and market opportunities of goat milk. Yangilar (2013) notes that the origin of this misconception is due to poor sanitary conditions in which goats are milked and that goat milk products are poorly manufactured.

The econometric results show that cost of disease control and number of education years attained were the significant socio-economic factors influencing goat milk production. Cost of labor inputs and number of does were insignificant. The explanatory power of the model was 0.606 which means that cost of disease control and number of education years attained explains 60.6% of the variation of goat milk output. The explanatory power of the model can be characterized as strong because it is closer to 1. The other 39.4% of the variation in goat milk output is explained by other factors which were not included in the model.

Many empirical studies have shown that cost of disease control and number of education years attained are significant factors that affect goat milk production. Kipresem et al (2011) in a similar study carried out in Kenya also reported similar results.

Ogola et al (2010) also found out that in Kenya, dairy goat farmers were not following recommended routine disease management practices due to high costs associated with drugs which affected dairy goat farming. These findings can be attributed to the fact that as farmer invests in disease control, diseases affecting goats are reduced thereby increasing milk production because healthy animals are more productive than sick animals.

Cost of disease control was the most significant and potent contributor to the total milk output in the smallholder areas of Zimbabwe (β=0.006, t=5.208, P<0.05). There is a positive relationship between cost of disease control and volume of milk produced with a Beta coefficient of 0.006(Table 5.5). This means that a \$1 increase in the cost of disease control per annum would lead to an increase of goat milk output by 0.006 litres per day. This shows an impact of an extra cost of disease control per annum on milk production. For the variable of cost of disease control per annum, the probability of the t-statistic (5.208) for the Beta coefficient is 0.000 which is significant at 5% level of significance. As expected, results indicate that there is a positive relationship between cost of disease control and volume of milk produced.

It was also noted that number of education years attained by the farmer is also a significant factor influencing goat milk production. This is in agreement with other studies which also showed that number of education years attained by farmer affected dairy goat farming. Ogola et al, (2010) found out that in Kenya the level of education attained significantly affected milk production. Number of education years attained (β=0.14; t=2.075; P<0.05) influences goat milk production in the smallholder areas of Zimbabwe. This means that a unit increase in the number of education years attained by household head would lead to an increase of goat milk output by 0.14 litres. The positive sign of the beta coefficient agrees with the hypothesized apriori sign. The probability of the t-statistic (2.075) in Table 5.5 for the Beta coefficient is 0.048 which is significant at 5% level of significance.

Results also show that cost of labor inputs and number of does are not significant factors that influence goat milk production in the smallholder areas of Zimbabwe. These findings are in disagreement with the research findings obtained in Keiyo North and Keiyo South districts of Kenya. Kipresem et al (2011) found out that cost of labor inputs and number of dairy goat stock are significant factors affecting dairy goat farming in Kenya. The insignificant effect of number of does and cost of labor inputs on goat milk production in the smallholder areas of Zimbabwe can be explained to be a result of the study limitations. Due to financial constraints the sample of respondents that was used was probably not big enough in order to detect significant differences. Though statistically insignificant, the sign of the coefficient cost of labor (0.004) inputs show the expected positive effect on goat milk output. This shows that an increase in labor inputs can lead to an increase in milk output. The sign of the beta coefficient of the number of does (-0.09) was different from the apriori expectations and maybe this was caused by the limitations of the study. Perhaps the sample of respondents that was used was probably not big enough in order to detect significant differences in some cases.

5.8 Summary of findings
The majority of the household heads (90%) who keep goats in Bulilima East District are literate. This means that communication of technical knowledge on goat milk production will be easy. The average number of goats per farmer was ten. There was no significant difference in costs of labor inputs and cost of disease control between farmers milking their goats and those who do not milk goats. Results have shown that only 5% of the farmers keep goats primarily for milk production. The primary reason for goat production is for cash income (73%). Results show that the main reason for not milking goats is because 50% of the households do not consume goat milk due to tastes and preferences. The econometric analysis of the socio-economic factors

influencing goat milk production has shown that cost of disease control and number of education years attained were significant. Cost of labor inputs and number of does were not significant variables. The R^2_{adj} of the model was 0.606 which means that 60.6% of the variation in goat milk output is explained by cost of disease control and number of education years attained. The other 39.4% of the variation in goat milk output is explained by other factors which were not in the model.

CHAPTER 6: CONCLUSIONS AND RECOMMENDATIONS

6.1 Introduction

The purpose of this chapter is to present the principle findings of the study, draw conclusions and policy recommendations. Areas for further research are also explored in this chapter.

6.2 Key research findings

Results of the study show that goat populations in Zimbabwe have been steadily increasing over the last two years. Cattle production in the country has been declining and this is an important opportunity for the goat industry since goat meat is a substitute of beef. About 97% of the goats belong to smallholder farmers and the majority of the goats (40.7%) are found in Matebelaland Provinces because of the favorable temperatures for goat production. Goat milk production is mainly undertaken in the Matebelaland Provinces.

The majority of smallholder farmers have smaller herd sizes of less than 8 goats and they depend on natural rangelands without supplementing their feeds. Shortage of feeds, higher mortality rates and the presence of an underdeveloped commercial market are some of the major challenges being faced by the farmers.

Evidence from the analysis of household survey data has shown that the main reason for goat production in the smallholder areas of Zimbabwe is income generation (72.5%) followed by meat (22.5%) and then lastly milk production (5%). The average volume of milk produced by a doe is 0.29 litres per day. Results also show that 50% of the households do not milk their goats because they do not consume goat milk due to taste and preferences. The production of little quantities of goat milk was also found to be the other reason why 45 % of the households do not milk their goats. The other 5% of the households do no milk their goats because of cultural reasons.

The ordinary least squares regression results show that cost of disease control and number of education years attained were the significant variables. Cost of labor inputs and number of does were insignificant variables. The explanatory power of the econometric model was 0.606 which means that cost of disease control and number of education years attained explains 60.6% of the variation of goat milk output. The explanatory power of the model can be characterized as strong because it is closer to 1. The other 39.4% of the variation in goat milk output is explained by other factors which were not included in the model.

6.3 Conclusions

Although goats in the smallholder areas of Zimbabwe are able to complement the value of cattle, this opportunity is being underutilized. Shortage of feeds, high goat mortality rates and the absence of a vibrant commercial market for goats are the major factors hindering farmers from getting the maximum benefits from their goats.

Goats in the smallholder areas of Zimbabwe are producing below the expected volume of milk. This is caused by the fact that the goats are not kept primarily for milk production. In addition, cost of disease control and number of education years attained by farmer are the main factors that are influencing goat milk production.

6.4 Policy recommendations

Since goats have not been kept primarily for goat milk production, the first strategy that can be used to improve goat milk production is to introduce dairy goat breeds in the smallholder areas. If smallholder farmers are assisted to get dairy breeds which are specifically for milk production, this will result in increased milk production. Dairy goat breeds produce more milk as compared to the non-dairy breeds in the smallholder areas of Zimbabwe.

Health care management strategies must be put in place by the government to increase goat milk production since cost of disease control has been found to be influencing goat milk production. Many farmers are not able to buy drugs for their goats and this is exposing the goats to diseases which lead to increased mortalities. Sick does will not be able to produce more milk due to poor health especially considering the fact that farmers are not able to buy drugs due to high costs.

The government must therefore promote and develop para-veterinary programs in the smallholder areas to improve access to health services. This will allow farmers access to drugs and veterinary services due to proximity. This reduces the costs of drugs and veterinary services, thereby making them affordable and accessible to smallholder farmers. In conjunction with the accessibility to improved breeds, this can overall lead to increased goat milk production.

6.5 Areas for further research

Further studies need to be conducted to determine on the best criteria for implementing specialized dairy goat production and how to improve the commercial markets of goats. Studies are also needed to assess the consumption pattern of goat milk and the contribution of goat milk to the livelihoods of smallholder farmers.

More research is needed on innovative cost effective ways and channels of communication and education tailored to cater for low-educated and illiterate goat farmers. A lot of work has to been done to evaluate the effectiveness of traditional remedies for disease control and their dosing rates. The key areas deserving more attention relates to the overall impact on household welfare, such as whether goat milking improves the food security of smallholder farmers. A broad range of attributes in milk desired by consumers should also be considered in future studies to explain the difference in preference between goat milk and cow milk.

Further studies must be conducted to determine the effect of other socio-economic factors on goat milk production. Effects of amount of feeds used, goat breed and number of extension visits on goat milk production must be evaluated. Number of extension visits was left out from the study to avoid getting biased answers since the data collection process was done in the presence of extension workers.

REFERENCES

Allam, M. R., 2000. Goat raising in the smallholder farming systems in Bangladesh. In: Proceedings of the Seventh International Conference on Goats, 15-21 May, 2000, Tours, France, pp. 329-330.

Armagan, G., Ozden, A., 2007. Determinations of total factor productivity with Cobb-Douglas production function in agriculture: The case of Aydin-Turkey. J. Appl. Sci., 7: 499-502.

Ayalew, W., Rischkowsky, B., King, J. M., Bruns, E., 2003. Crossbreds did not generate more benefits than indigenous goats in Ethiopian smallholdings. Agricultural Systems 76: 1137-1156.

Ayele, Z., Peacock, C., 2003. Improving access to consumption of animal source foods in rural households: the experiences of a women- focused goat development program in the highlands of Ethiopia.J.Nutr.133 (11): 39815-39865.
Boor K J, Brown D L and Fitzhugh H A 1987 Western Kenya: The potential for goat milk production. World Animal Review 62 (1987) 31-40

Cannas, A., Pulina, G., 2005. Dairy goats: feeding and Nutrition. Department of Animal Science. University of Sassari, Italy.

Chifamba, I. K., Khombe, C.J., Sithole, L., 1993.Comparative performance of indigenous Boer goats under Brachysegial Julbernadia woodland, in Small ruminant production in Zimbabwe: Prospects and constraints: Proceedings of a workshop at Matopos Research Station, 19-20 August 1993, edited by L. M. Sibanda and supported by the University of Zimbabwe, Ministry of Lands, Agriculture and Water Development and the French Embassy, Zimbabwe: 28-33.

Chimowa, A., 2008.Farm Management Handbook: Section B. Livestock Production, Department of Agricultural and Extension Services, Zimbabwe.

De A.J.BOER., Fitzhugh, H.A., 1983.Sheep and goats in developing countries, their present and potential role.

Donkin, E.F., Boyazoglu, P.A., 2000. Milk production from goats for households and small-scale farmers in South Africa, In: Proceedings of the Seventh International Conference on Goats, 15-21 May, 2000, Tours, France, pp. 324-326.

Dubeuf, J. P., Morand-Fehr, P., Rubino, R., 2004.Situation, changes and future of goat industry around the world. Small ruminant Research 51 (2004) 165-173.

Eik, L.O., 1991. Effects of feeding intensity during dry period on performance of dairy goats, *Journal of Animal Science, Vol 6, pages 223-232.*

FAO., 1991. Production Year book, Food and Agriculture Organization of the United Nations, FAO, Rome Italy , Statistics Series No:140;45;265

FAO., 2005. Livestock Sector Brief Zimbabwe. Rome, Italy: Food and Agriculture Organization.

FAOSTAT 2011 Food and Agriculture Organization of the United Nations FAO Statistics Available at http://faostat.fao.org/site/573/DesktopDefault.aspx?PageID=573#ancor.

Flatten, O., 2012. Profitability of early grass silage harvesting on dairy goat farms in mountainous areas of Norway, small ruminant research , *Journal of Animal Science, Vol 103,Issue 2-3, pages 133-142.*

Gihad, E. A., El-Bredawy, T. M., 2000. Contribution of goats to Egyptian small farmer food and income in three systems. In: Proceedings of the Seventh International conference on goats, 15-21 May, 2000, Tours, France, pp 531-534.

Harlan, H., Atfield, D., 2003.Understanding Goat Production .Agricultural Research Service, United States, Department of Agriculture.

Homann, S., Rooyen, A., 2006.Goats in semi-arid Zimbabwe. The importance and the development challenge.

Homann, S., van Rooyen, A., Moyo, T., Nengomasha, Z., 2007. Goat production and marketing: Baseline information for semi-arid Zimbabwe, ICRISAT, Bulawayo, Zimbabwe.

Hove, T., Lind, P., Mukaratirwa,S., 2005. Seroprevalence of Toxoplasma gondii infection in goats and sheep in Zimbabwe.*Onderstepoort Journal of Veterinary Research, 72:267-272(2005).*

Khanna, M., 2011. Improving income-livestock distribution, ZimPro/ LDS farming project, Matabeleland South, Zimbabwe.

Kipresem, J. Sulo,T., Tuitoek, D., Saina, E., 2012. Production allocation efficiency of dairy goat farming in Elgeyo-Marakwet country in Kenya. *International Journal of Strategic Management, International Academy of Business and Economics. Volume 12, Issue 1.*

Kipserem, J., Sulo, T., Chepngeno, W., Korir, M., 2011. Analysis of factors affecting goat faming in Keiyo North and Keiyo South Districts of Kenya, *Journal of Developmental and Agricultural Economics, Vol 3(11), pages 556-560.*

Kosgey, I. S., Baker, R. L., Udo, H. M. J., van Arendonk, J. A. M ., 2006. Successes and failures of small ruminant breeding programmes in the tropics: a review. Small Ruminant Research 61: 13-28.

Lindela, R., Ndlovu, O., Matika., 1991. Goat development in Zimbabwe, Prospects and constraints, Matopos Research Station.

Manyenga A 1987 Acceptability of goat milk. Farming World September 1987. pp23.

Ministry of Agriculture Mechanization and Irrigation Development., 2013. Zimbabwe.

Morand-Fehr, P., 1991.Goat Nutrition EAAP Publication No. 4623.

Morand-Fehr, P., Bas, P., Blanchart, G., Daccord, R., Giger-Reverdin, S., Gihad, E.A., Hadjipanayiotoui, M., Mowlem, A., Remeuf, F., Sauvant, D., 1991.Influence of feeding on goat milk composition and technological characteristics, Goat nutrition. EAAP Publication No.46.

Morand-Fehr, P., 1980.Composition and yield of goat milk as affected by nutritional manipulation, *Journal of Dairy Science, Vol 63,Issue 10, pages 1671-80.*

Mowlem A 2005 Marketing dairy goat produce in the UK. Small Ruminant Research 60 (1) 207-213

Murithi, F.M., 1990. Efficiency of Resource use in Smallholder Milk Production. The case of Meru Central Dairy Farmers, Kenya. Unpublished M.Sc Thesis, University of Nairobi.

Nicholas, N., 2007. Enhancing incomes and livelihoods through improved farmers' practices on goat production and marketing, Context of the Goat Sector in Zimbabwe.

Ndlovu, L.R., 1991. Goat Development in Zimbabwe: Prospects and Constraints. Proceedings of a workshop held at Matopos research station, 14-15 November 1991.

Nyathi, N., Moyo, E., Sarupinda, D., 2007. Sub-sector analysis report. Unlocking economic opportunities for smallholder goat producers through the promotion of access to markets. Bulawayo, Zimbabwe: The Netherlands Development Organization.

Ogola, T. D. O., Nguyo, W .K., Kosgey, I. S., 2010. Economic contribution and viability of dairy goats: implications for a breeding programme. Tropical Animal Health and Production 42: 875-885.

Ogola, T., Nguyo, W., Kosgey, I., 2010. Dairy goat production practices in Kenya: Implications for a breeding program, Livestock Research for Rural Development 22(1), 2010.

Onim, J .F. M., 1992., Dual-purpose goat research in western Kenya. In: Kategile, J.A., Mubi, S. (editors), Future of livestock industries in East and Southern Africa, Proceedings of a Workshop held at Kadoma Ranch Hotel, Zimbabwe, 20-23 July, 1992. ILCA (International Livestock Centre for Africa), Addis Ababa, Ethiopia, 227 pp. 5.

Onim, JFM., 2009. Evaluation and Effectiveness of breeding and production services for dairy goat farmers in Kenya, Journal of Dairy Science, Vol 103, Issue: 8-9, pages 223-232.

Peacock, C., 1991. Improving goat production in the tropics, A manual for development workers, Oxfam publications.

Peacock, C., 2005. Goats unlocking their potential for Africa's farmers. In: Proceedings of the Seventh Conference of Ministers Responsible for Animal Resources, 31 October – 4 November, 2005, Kigali, Rwanda, pp. 1-23.

Peacock, C., 1996. Improving Goat Production in the Tropics. A Manual for Development Workers, Oxfam Publications.

Riviere,J.,1993. Small ruminant production in Zimbabwe; Prospects and Constraints. Proceddings of a workshop held at Matopos 19-20 August ,1993.

Rukuni, M., Eicher, C., 2006. Livestock research and development, Zimbabwe Agricultural Revolution, 2nd Edition, UZ Publications, Harare.

Rumosa, G., Chimonyo, M., Dzama, K., 2008. Communal goat production in Southern Africa: *A review Journal of Tropical animal health and production, Vol 41, number 7, pages 1157-1168.*

Sabine, H., Andre, R., Thinah, M., Zivayi, N., 2007. Goat production and marketing, Baseline information for semi arid Zimbabwe, Matopos Research Station.

Sarupinda, D., Tavesure, F., 2009.Goat Auctions opening market opportunities for smallholder farmers in western provinces of Zimbabwe.

Shah, A., Saboor. A., Ahmad. S., 2009. An estimation of cost of milk production in Pakistan: A microeconomic approach. Sarhad J. Agric. 25(1): 141-147.

Sibanda, L.M., 1993. Sensitivity analysis as a tool for ranking constraints in a smallholder production system in Small ruminant production in Zimbabwe: Prospects and constraints: Proceedings of a workshop at Matopos Research Station, 19-20 August 1993, edited by L. M. Sibanda and supported by the University of Zimbabwe, Ministry of Lands, Agriculture and Water Development and the French Embassy, Zimbabwe: 20-27.

Sibanda, R., 2005. Livestock development in Southern Africa: Future research and investment priorities. Zimbabwe country report. International Crops Research Institute for the Semi-Arid tropics, Bulawayo, Zimbabwe.

Sibanda S and Khombe C T 2006 Livestock research and development. In: Rukuni M, Eicher C, with Mabel Munyuki-Hungwe and Matondi P (editors) 2006 Zimbabwe's Agricultural Revolution Revisited. University of Zimbabwe Publications, Harare, Zimbabwe.

Sikosana, J.N., 2007. Enhancing incomes and livelihoods through improved farmers' practices on goat production and marketing. Proceedings of a workshop organized by the Goat Forum, Bulawayo, Zimbabwe, 2–3 October 2007.

Siringwani, H., Jiri, 0., 2010. Practical Agricultural Experience, AGPR 102 and 103 courses. Student handbook. University of Zimbabwe.

Steele, M., 1996.The tropical agriculturalist. Centre for Tropical Veterinary Medicine, University of Edinburgh.

Torane, S. R., 2009.An econometric analysis of farming Systems In North Konkan Region Of Maharashtra. PHD Thesis, University of Agricultural Sciences, Dharwad.

Van Raust, G.V. Fierest,J.,De Rick, Van Bockstack, E., 2009. Influence of ensiling forages at different dairy matters and silage additives on lipid metabolism and fatty acids composition. Animal feed science technology. 150(1-2): 62-74.

Van Rooyen, A., Homann, S., 2009. Promoting goat marketing and technology development in Semi-arid Zimbabwe for food security and income growth; Tropical and Sub-tropical Agro-ecosystem Vol 11, Num.1, pp 1-5.

van Rooyen, A., Homann,S., 2007.Enhancing incomes and livelihoods through improved farmers' practices on goat production and marketing, Proceedings of a workshop organized by the Goat Forum, Bulawayo, Zimbabwe.

Yangilar F 2013 As a potentially functional food: Goat's milk and products. Journal of Nutrition Research 1(4) 68-81

APPENDIX 1: National estimates and distribution of goats in Zimbabwe by province in 2012

Province	Goat population
Manicaland	634742
Mashonaland central	227751
Mashonaland east	154502
Mashonaland west	85546
Matabelaland north	805884
Matabelaland south	445827
Masvingo	289180
Midlands	429418
Total	3072850

Source: Ministry of Agriculture, Mechanisation and Irrigation Development (2013)

APPENDIX 2: Household survey questionnaire

Socio-economic factors influencing goat milk production in the smallholder areas of Zimbabwe: Case study of Bulilima district.

SECTION A: IDENTIFICATION INFORMATION

Name of Enumerator...

Name of Respondent...

Village Name...

Ward..

SECTION B HOUSEHOLD DEMOGRAPHIC DATA

B1. Who is the household head?

1. Father   **2.** Mother   **3**. Child   **4.** Others (specify).....................

B2 Age of household head..................................

B3 What is the number of household members?..

B4 . What is the highest level of education attained by the household head?

1.None   **2.** Grade 7   **3**. ZJC   **4**. Ordinary level   **5**. Advanced level   **6**.Tertiary

B5. Do you hold any agricultural qualifications?   **1.** Yes   **2.**   No

B6. If yes what type of qualification?   **1**. Master Farmer   **2**. Agricultural certificate

3. Agricultural diploma   **4**. Others specify ....

SECTION C: GOAT OWNERSHIP

C1. How many goats do you keep?..

C2. Indicate the composition of your flock.

Mature goats: males [   ]   females [   ]

         Kids: males [   ]   females [   ]

C3. What is the main reason for keeping goats?

1.Income   **2**. meat   **3**.milk   **4**. others (specify)...............

C4. When did you start rearing goats?...

C5. Do you milk your goats?   **1** .Yes   **2**. No

If your answer above is **NO** go to question **C8.**

C6. If your answer above is yes how many times a day do you milk your goats? ....

C7 .What is the volume of milk in litres obtained from a goat per day..

C8. What is the reason for not milking your goats?

1. We don't consume the milk   **2.** Very little quantities of milk are produced

 3. others (specify).................................

C9. List according to importance five problems you encounter in rearing goats

1...

2...

3..

4..

5..

SECTION D: LABOR MANAGEMENT

D1. Do you use your own family labor for goat rearing?   1. Yes   2. No

D2. If yes, how is labor allocated within the family?

Activity	Person responsible	Reason
Milking		
Feeding		
Cleaning the stall		
Herding		
Vaccinating		

D3. Do you employ hired labor for goat rearing?   1. Yes   2. No

If your answer is **No** go to question **D6**

D4. How many people do you employ to assist you in rearing goats?...

D5 .How much do you pay the hired labor per month?...

D6. If you were to use hired labor, how much would you be willing to pay a single worker per month?..

SECTION E HEALTH CARE MANAGEMENT

E1. What measures do you take when your goats are affected by diseases?

1.treat them   **2.** Take them to the veterinary clinic   **3**. slaughter them
4. sell them immediately   **5**. do nothing   6. Others specify..........................

E2.Do you use any medicines or equipment for treating goats?   **1** Yes   **2** No

If yes, what type of medicines do you use and where do you buy them and their costs?

Type of medicines	Where do you buy them	Transport costs	Costs of medicine per year

E3. What are the challenges that you encounter in providing health care and what measures do you take to overcome them?

Challenges in providing health care	Measures taken

Health care management practices (tick where appropriate)

Practice	Indicate by a tick	Frequency per year	Drug used	Cost per year
Dipping				
Dosing				
Vaccination				

E4. How many kids died in the last year?...

E5. How do you deal with animals that die due to diseases?...

E6 What are the most common causes of mortality?

1.predators   **2** diseases   **3.** Accident   **4.** Drought   **5.** Posoining   **6** unknown

7.others specify..

E7.Do you vaccinate your goats?   1.Yes   2.No

E8. If your answer is no what is the main reason for not vaccinating your goats?

1 . Lack of know-how   **2.** it's expensive   **3.** Shortage of veterinarians

4. others(specify)...

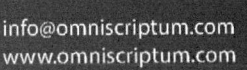